ANIMALS OF THE FOREST

Red Foxes

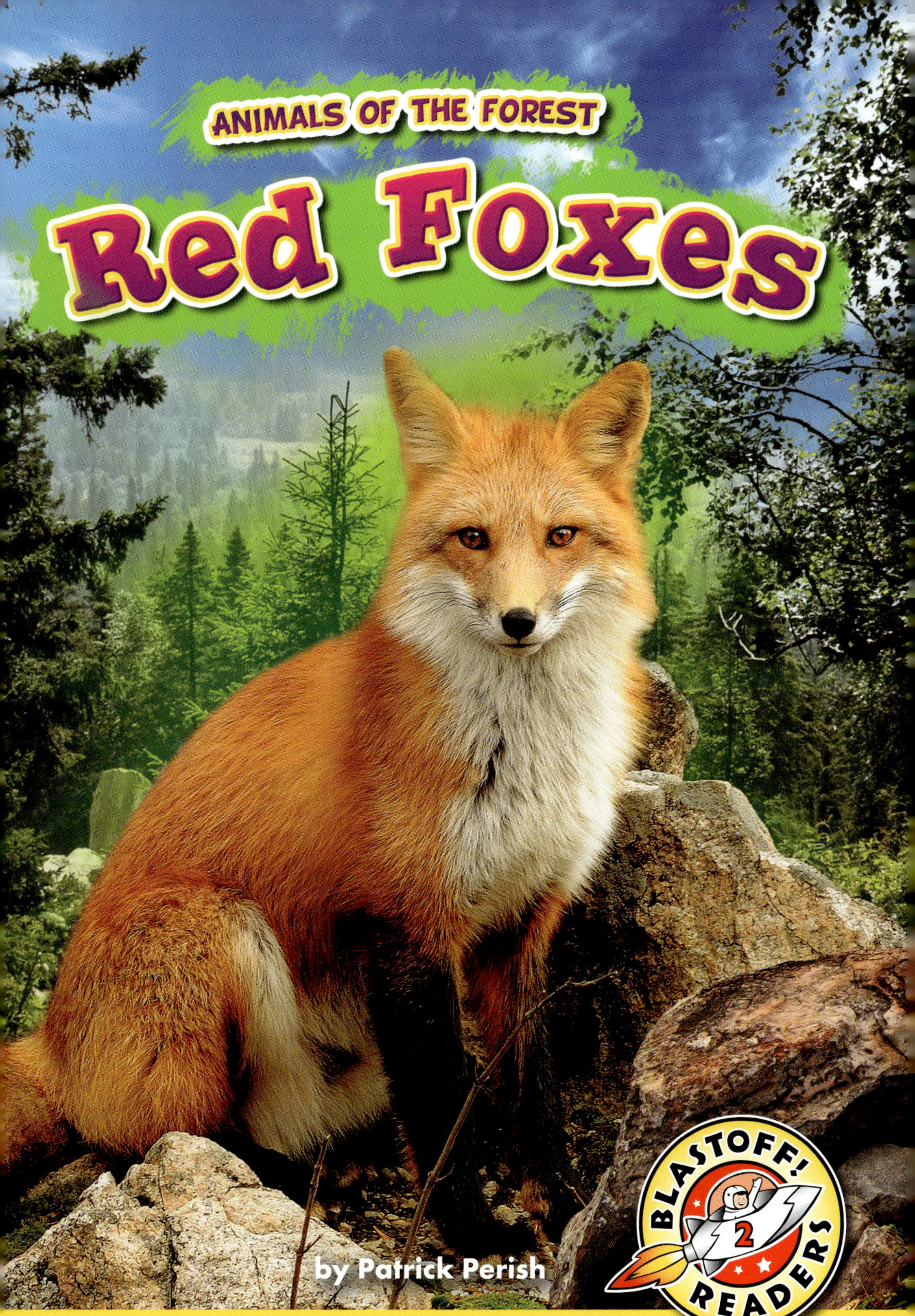

by Patrick Perish

BELLWETHER MEDIA • MINNEAPOLIS, MN

Blastoff! Readers are carefully developed by literacy experts to build reading stamina and move students toward fluency by combining standards-based content with developmentally appropriate text.

Level 1 provides the most support through repetition of high-frequency words, light text, predictable sentence patterns, and strong visual support.

Level 2 offers early readers a bit more challenge through varied sentences, increased text load, and text-supportive special features.

Level 3 advances early-fluent readers toward fluency through increased text load, less reliance on photos, advancing concepts, longer sentences, and more complex special features.

★ Blastoff! Universe

This edition first published in 2022 by Bellwether Media, Inc.

Library of Congress Cataloging-in-Publication Data

Names: Perish, Patrick, author.
Title: Red foxes / by Patrick Perish.
Description: Minneapolis, MN : Bellwether Media, [2022] | Series: Blastoff! readers: Animals of the forest | Includes bibliographical references and index. | Audience: Ages 5-8 | Audience: Grades 2-3 | Summary: "Relevant images match informative text in this introduction to red foxes. Intended for students in kindergarten through third grade"– Provided by publisher.
Identifiers: LCCN 2021004023 (print) | LCCN 2021004024 (ebook) | ISBN 9781644875087 (library binding) | ISBN 9781648344169 (ebook)
Subjects: LCSH: Red fox–Juvenile literature.
Classification: LCC QL737.C22 P438 2022 (print) | LCC QL737.C22 (ebook) | DDC 599.775–dc23
LC record available at https://lccn.loc.gov/2021004023
LC ebook record available at https://lccn.loc.gov/2021004024

Editor: Kieran Downs Designer: Josh Brink

Printed in the United States of America, North Mankato, MN.

Table of Contents

Life in the Forest

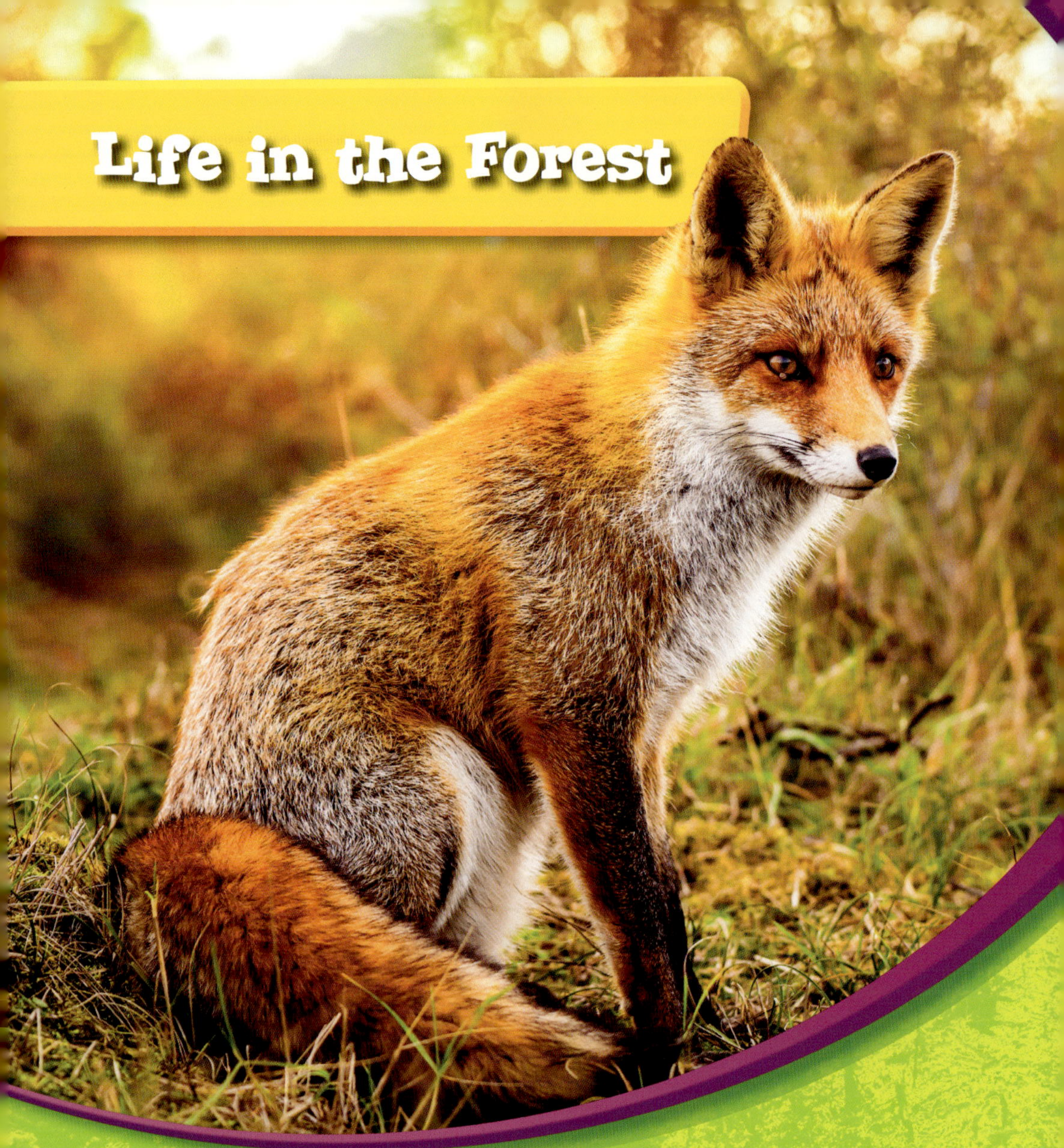

Red foxes are found throughout the northern half of the world. They also live in Australia.

They have **adapted** to the forest as well as many other **biomes**.

Red Fox Range

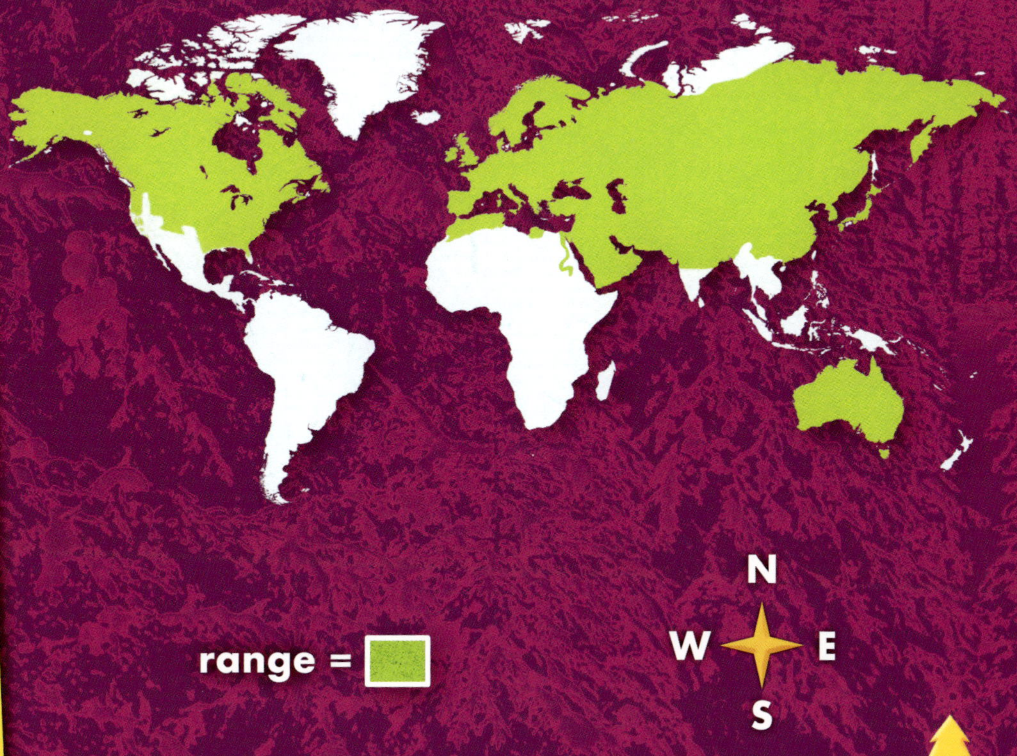

Forests can be difficult places to hunt. Red foxes have sharp senses to find food.

Their eyes see well in low light. Their large ears hear **prey** underground.

Forest prey can be quick. But red foxes are quick **predators**.

Their long legs help them run, jump, and chase prey.

Red foxes have big, **bushy** tails. These help foxes balance as they chase prey.

Forest winters can get cold. Foxes use their tails as blankets when they sleep.

Special Adaptations

Fox Homes

Red foxes often live in **dens**. They store the food they find in the forest there.

Tunnels connect the dens to places where the foxes hunt.

den

Red foxes mostly live alone. They mark their **territories** by putting their **scents** on logs and rocks.

These smells tell other foxes to hunt somewhere else.

Forest animals are often most active at night. This is when most red foxes hunt.

Red foxes often rest during the day.

Red Fox Stats

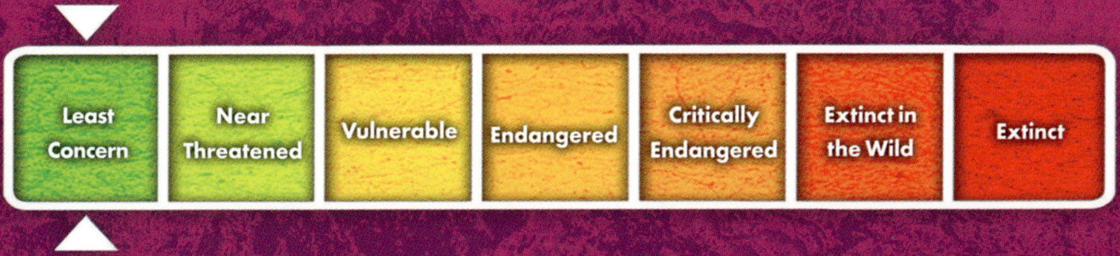

conservation status: least concern

life span: up to 4 years

Fox Food

Red foxes are not picky eaters! These **omnivores** eat mice, birds, bugs, and berries.

They even eat **carrion** found in the forest.

Forest Diet

deer mouse

lowbush blueberry

ruffed grouse

Red foxes **pounce** on their prey and bite it. They bury extra food to eat later.

These smart animals **thrive** in their forest homes!

Glossary

adapted—changed over a long period of time

biomes—large areas with certain plants, animals, and weather

bushy—thick and furry

carrion—the rotting meat of a dead animal

dens—sheltered places

omnivores—animals that eat both plants and animals

pounce—to suddenly jump on something to catch it

predators—animals that hunt other animals for food

prey—animals that are hunted by other animals for food

scents—smells that show where an animal has been

territories—land areas where animals live

thrive—to grow well

To Learn More

AT THE LIBRARY

London, Martha. *Foxes.* Minneapolis, Minn.: ABDO, 2021.

Perish, Patrick. *Fennec Foxes.* Minneapolis, Minn.: Bellwether Media, 2019.

Pettiford, Rebecca. *Arctic Foxes.* Minneapolis, Minn.: Bellwether Media, 2019.

ON THE WEB

FACTSURFER

Factsurfer.com gives you a safe, fun way to find more information.

1. Go to www.factsurfer.com.
2. Enter "red foxes" into the search box and click 🔍.
3. Select your book cover to see a list of related content.

Index

The images in this book are reproduced through the courtesy of: Rejean Bedard, cover, p. 12; Rvkamalov gmail.com, cover (background); Alex Stemmer, CIP; SanderMeertinsPhotography, p. 4; WildMedia, pp. 6, 7; Pim Leijen, p. 8; Milan Zygmunt, p. 9; Menno Schaefer, p. 10; Mark Bridger, p. 11 (large ears); Holly S Cannon, p. 11; Geoffrey Kuchera, pp. 12-13; David Kalosson, p. 14; davemhuntphotography, p. 15; AB photographie pp. 16, 23; Dgwildlife, p. 17; Vaclav Sebek, p. 18; jitkagold, p. 19 (deer mouse); Mircea Costina, p. 19 (lowbush blueberry); LarryDallarire, p. 19 (ruffed grouse); Mal Harris, p. 20; Cezary Wojtkowski, p. 21.